baby GECKOS

KIM THOMPSON

CREATIVE EDUCATION • CREATIVE PAPERBACKS

CONT

ENTS

I AM A HATCHLING.

I am a baby gecko.

mouth
My feet are covered with tiny hairs. They stick on anything.
tiny hairs
foot

My mom laid an egg under leaves. In a few months, I hatched.

I do not need an adult to care for me. I can survive on my own.

My long tongue catches insects. My sticky feet climb trees. My colors help me hide.

I stay awake all night. I sleep all day.

Snakes, birds, and other animals want to eat me.

If an animal gets my tail, it breaks off. This does not bother me. I can grow a new tail!

I am one of the smallest lizards.

SPEAK AND LISTEN

TUH-K

Can you speak like a hatchling?

Baby geckos chirp and grunt.

Listen to these sounds:

https://www.youtube.com/watch?v=03eS6Ph91EI&t=75s

GECKO WORDS

hatched: broke out of an egg to be born

insects: small animals with six legs and hard outer skeletons; bugs

lizards: reptiles that have scaly bodies, four legs, and long tails

scales: thin, flat, overlapping pieces of hard skin that cover the bodies of lizards and other reptiles

READING CORNER

Huber, Raymond. *Gecko*. Somerville, Mass.: Candlewick Press, 2019.

McCarthy, Colin. *Eyewitness Reptile*. London: DK Children, 2023.

Riggs, Kate. *Geckos (Amazing Animals)*. Mankato, Minn.: Creative Paperbacks, 2022.

INDEX

PUBLISHED BY CREATIVE EDUCATION AND CREATIVE PAPERBACKS
P.O. Box 227, Mankato, Minnesota 56002
Creative Education and Creative Paperbacks are imprints of The Creative Company
www.thecreativecompany.us

LIBRARY OF CONGRESS CATALOGING-IN-PUBLICATION DATA
Names: Thompson, Kim, 1970- author
Title: Baby geckos / Kim Thompson.
Description: Mankato, Minnesota : Creative Education and Creative Paperbacks, [2026] | Series: Starting out | Includes bibliographical references and index. | Audience term: juvenile | Audience: Ages 4-7
Creative Education and Creative Paperbacks | Audience: Grades K-1 Creative Education and Creative Paperbacks | Summary: "Introduce beginning readers to the world of baby geckos with this life science starter. Includes photos, a labeled animal diagram, "Make a Noise" section, glossary, and further resources"-- Provided by publisher.
Identifiers: LCCN 2024043239 (print) | LCCN 2024043240 (ebook) | ISBN 9798889897477 library binding | ISBN 9781682778333 paperback | ISBN 9798889897606 ebook
Subjects: LCSH: Geckos--Juvenile literature. | Animals--Infancy--Juvenile literature
Classification: LCC QL666.L245 T46 2026 (print) | LCC QL666.L245 (ebook) | DDC 597.95/2--dc23/eng/20250107
LC record available at https://lccn.loc.gov/2024043239
LC ebook record available at https://lccn.loc.gov/2024043240

DESIGN AND PRODUCTION
Design by Rhea Magaro
Production by Beeline Media and Design, Inc.
Art direction by Tom Morgan

PHOTOGRAPHS by Alamy Stock Photo/Alon Meir, 7, Chris Mattison, 6-7, FB-StockPhoto-1, 5; Shutterstock/Agus_Gatam, cover, 4, 14, Bohbeh, 13, DWI YULIANTO, 9, Kurit afshen, 2-3, 8, Matt Jeppson, 10-11, Robert Eastman, 11, somsak mungmee, 12

Printed in India